A Question of Math Book

Subtraction

by Sheila Cato
illustrations by Sami Sweeten

Carolrhoda Books, Inc./Minneapolis

This edition published in 1999 by Carolrhoda Books, Inc.

Carolrhoda Books, Inc., c/o The Lerner Publishing Group
241 First Avenue North, Minneapolis, MN 55401 U.S.A.

Website address: www. lernerbooks.com

LIBRARY OF CONGRESS CATALOGING-IN-PUBLICATION DATA
Cato, Sheila
    Subtraction / by Sheila Cato : illustrations by Sami Sweeten.
        p.   cm. — (A question of math book)
    Summary: A group of children introduce subtraction, using everyday examples and practice problems.
    ISBN 1-57505-318-7 (alk. paper)
    1. Subtraction—Juvenile literature.   [1. Subtraction.]
I. Sweeten, Sami, ill. II. Title. III. Series: Cato, Sheila, 1936-
Question of math book.
QA115.C28 1999
513.2'12—dc21   98-6355

The series A Question of Math is produced by Carolrhoda Books, Inc., in cooperation with Brown Packaging Partworks Limited, London, England. The series is based on a concept by Sidney Rosen, Ph.D.
Series consultant: Kimi Hosoume, University of California at Berkeley
Editor: Anne O'Daly
Designers: Janelle Barker and Duncan Brown

Printed in Singapore
Bound in the United States of America

1 2 3 4 5 6 - JR - 04 03 02 01 00 99

eet Holly. She is learning about subtraction with the help of her number friend, Digit. Holly's other friends will be joining in the fun, too. There are also some easy subtraction problems for you to try. You can use buttons, beads, pennies, and marbles to help you figure them out.

My little cousin Luke has come to visit. I have 7 toys and I am going to give him 3 of them. How many toys will I have left?

Taking away is called subtraction.
Let Luke take the toys he wants.
Now count how many toys you have left.

4

I started with 7 toys. Luke took 3, and I have 4 toys left.

You can write about this in a special way

$$7 - 3 = 4$$

This is called an equation.

Special Signs

Take a look at the special signs in Holly's equation. These signs tell us what to do. The sign - means take away, minus, or subtract. The sign = means equals or is the same as.

Now You Try

If Holly started with 7 toys and gave Luke 4 of them, how many toys would she have left? Use buttons to help you.

Josh and I are baking a cake. There are 10 eggs in the bowl, and we're going to use 5 of them. My mom wants to make an omelette with the eggs we don't use. How many eggs will be left?

You can work this out easily, Holly. Your hands are a group of 10 fingers. Each hand is a group of 5. So what happens when you take a group of 5 away from a group of 10?

3 toys

Number Sentence

When we speak or write, we put the words into sentences. When we write about math like this – "Seven minus three equals four" – it is called a number sentence.

10 take away 5 leaves 5. If we use 5 eggs to bake the cake, there'll be 5 eggs left over for the omelette.

$$10 - 5 = 5$$

Now we just have to wait until the cake is cool enough to eat!

Now You Try

Write the number sentence that shows 10 - 5 = 5.

*I need to practice subtracting 2 at a time.
Can you think of an easy way to keep taking
2 away from 10?*

I sure can, Holly. Hold up your hands
with all your fingers pointing
upward. Now bend your thumbs
down. You have 8 fingers standing.
10 - 2 = 8.

Now bend the fingers next to your
thumbs. You have 4 fingers pointing
down and 6 fingers raised. 10 - 4 = 6.

Pairs

A pair is 2 of something. Lots of things come in pairs – hands, feet, socks, shoes, roller blades, earrings. It's good to know how to subtract in 2s.

If I bend the next 2 fingers, I get 10 - 6 = 4.
The next 2 fingers give 10 - 8 = 2.
When I bend my little fingers, there are no fingers still raised.

$$10 - 10 = 0$$

I can use my fingers to figure out lots of subtraction problems!

Now You Try

Cut out 5 pairs of things from magazines. Cover up 2 pairs. How many pairs are left? Can you write an equation for this?

I have 9 rabbits, but that's too many for me to take care of. I'm going to give 1 rabbit each to Brad, Mia, and Luis. How many rabbits will I have left?

Start at 9, and count down the numbers as you give each person a rabbit.
Give 1 rabbit to Luis and say "8."
Give 1 rabbit to Mia and say "7."

10 - 4 = 6

Sets and Groups

A set is a group of things. The members of the set can be anything you can think of – including toys, people, eggs, or rabbits. Subtraction is about taking a smaller set away from a bigger set.

Now give a rabbit to Brad and say "6."
That is the number of rabbits you have left.

That's an interesting way to take 3 away from 9. I have 6 rabbits left, and I can write this as an equation

$$9 - 3 = 6$$

It's going to be much easier to take good care of 6 rabbits.

11

Now You Try

Imagine Holly started with 6 rabbits. If she gave 3 rabbits away, how many would she have left?

Our favorite animal at the aquarium is the octopus, but today he won't come out from behind his rock. We know he has 8 arms, but we can only see 5 of them. How many arms are hidden behind the rock?

This is like asking "8 - ? = 5."
Find 8 buttons. Take away one button at a time until you have 5 left. How many buttons did you take away? This is the number of arms the octopus has hidden behind the rock.

6 - 3 = 3

I took 3 buttons away from the group of 8. The octopus has 3 arms hidden.

$$8 - 3 = 5$$

I hope the octopus is feeling more friendly next time we visit!

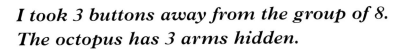

One Way

Subtraction is like a one-way street – it only works one way. Take a small set from a big one.
$8 - 2 = 6$.
But 2 minus 8 does not equal 6.

Now You Try

If the octopus only had 2 arms showing, how many would be hidden?

Here I am at the mall with Brad, Mia, and Luis. We love riding on the escalators. There are 8 escalators all together. 4 are going up. The other escalators are coming down. How many are coming down?

Okay, Holly, let's start with what you already know. There are 8 escalators. 4 are going up. Ride up to the next floor on the up escalators.

6 arms

Now you can ride back down on the down escalators and count as you get to the bottom.

1, 2, 3, 4! There are 4 up escalators and 4 down escalators.

$$8 - 4 = 4$$

Be careful when you ride on escalators!

Checking the Answer

Subtraction involves taking a small set away from a bigger set. To check your answer, add the set you take away to the set you have left. You should get the set you started with.

Now You Try

If the mall had 12 escalators, with 6 going up, how many would be coming down?

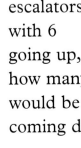

15

Here we are at the bowling alley. I just had my turn, and there are 3 pins left standing. I know there were 10 pins to start with. How many did I knock over?

Taking a number away from 10 is easy, because you have 10 fingers to help figure it out. Bend your fingers until 3 are left standing. How many fingers have you bent?

12 - 6 = 6

Opposites

Many words have opposites, like big and small, short and tall. Subtraction separates a big set into 2 smaller sets. Addition joins 2 smaller sets to make a big set. Addition is the opposite of subtraction.

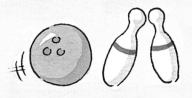

I've bent 7 fingers, so I know I knocked over 7 pins. That's one way to find the answer. But I could also have subtracted 3 from 10 to get 7.

$$10 - 3 = 7$$

Now that I know so much about subtraction, I can use different ways to figure out the problems. I can also check my answer to see if I'm right. 7 + 3 = 10.

Now You Try

If Luis knocked down 8 of the pins, how many would be left standing? Try to figure this out in two different ways and check your answer.

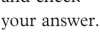

17

10 - 8 = 2, 10 - 2 = 8. Check: 2 + 8 = 10

My dad bought 10 rolls of wallpaper to decorate my bedroom. Now that he has finished, he has 4 rolls left. How many rolls did he use?

Your dad used a smaller number than he bought. If you take 4 away from 10, that will give you the answer.

Less Than

When a set or number is smaller than another set or number, we say it is less than the larger set or number. This has a special sign, too. The sign < means less than.

6 < 10.

I know that if I take 4 away from 10 I get 6. I already solved that problem on page 8.

$$10 - 4 = 6$$

My dad used 6 rolls of wallpaper. This is less than the number of rolls he bought.

Now You Try

Imagine that Holly's dad used 3 rolls less than he bought. How many rolls did he use?

Who will win the tug-of-war between the girls and the boys? I can see that there are more girls than boys, but how many more?

There are 9 girls and 6 boys. We can use one-to-one matching to find out how many more girls there are. Put 9 red counters in a line. Next find 6 blue counters and place one blue counter on top of a red counter until you run out.

He used 7 rolls

How many red counters are left without a blue counter? That's how many more girls there are than boys.

There are 3 red counters without a blue counter on top. So there are 3 more girls than boys. Another way of doing this would be to say

$$9 - 6 = 3$$

It looks like the girls have won!

How Many More?

We can subtract one set from another, to find out how many things there are in the larger set than the smaller one.

Now You Try

If there were 12 girls and 7 boys, how many more girls would there be?

Brad and I like to visit the baker. He is making cupcakes. There are 16 cupcakes on the tray, but only 12 have been frosted. How many still need to be frosted?

We can easily figure this out. Count down from 16 until you get to 12. Raise one finger as you say each number.

22

That sounds easy – 15, 14, 13, 12. I have 4 fingers raised, so 4 cupcakes still need to be frosted.

$$16 - 12 = 4$$

If we help the baker with the frosting, I wonder if he'll give us a cupcake each?

Even Numbers

All even numbers end in 0, 2, 4, 6, or 8. When you take one even number away from another even number, the answer is always an even number.

Now You Try

Suppose the baker made 14 cupcakes. If he had frosted 8, how many would need to be frosted?

Mia has come to visit with her dog, Popcorn. When Mia and I left the kitchen, there were 9 cookies on a plate. Now there are only 4. We know Popcorn ate the missing cookies, but how many did she take?

It looks like Popcorn has done some subtraction by herself! You can figure out the answer using counters.

Get 9 counters and separate them into two sets so that there are 4 counters in one set. How many counters are in the other set?

6 cupcakes

24

Odd Numbers

Odd numbers end in 1, 3, 5, 7, or 9. When you take an odd number away from an odd number, the answer is an even number. When you take an even number from an odd number, the answer is odd. When you take an odd number from an even number, the answer is odd.

There are 4 counters in one set and 5 counters in the other set. That means Popcorn took 5 cookies.

$$9 - 5 = 4$$

I hope she doesn't get sick!

Now You Try

Get 10 counters and separate them into 2 groups with an odd number of counters in each group. How many different ways can you do this?

*Josh is helping me paint our fence.
There are 16 fence boards, and we've
painted 10. How many more fence boards
need to be painted?*

Your fence is like a number line with the
gate post as 0. Start at the 16th board and
move back to the 10th one, counting the
boards as you go. That will tell you how many
boards you still need to paint.

1 and 9, 3 and 7, 5 and 5

When I move from 16 to 15, that's 1 space. From 15 to 14 is another space, so I've moved 2 spaces all together. If I keep going until I get to 10, I've moved 6 spaces.

$$16 - 6 = 10$$

So I have 6 more boards to paint.

Number Line

Counting down on a number line is a good way to figure out subtraction problems. You can make your own line by writing numbers on a piece of paper. A number line starts at 0 and goes on and on to whatever number you want!

Now You Try

Count back on the number line to see how many more boards need painting when Holly has painted 7 of them.

When Luis and I walk to school along the main road, it takes us 10 minutes. If we take a shortcut through the park, we get to school 3 minutes sooner. How long does the quick route take?

We can use subtraction to answer your question, Holly. It takes 3 minutes less to walk through the park than along the road. How would you figure this out?

28

Take Your Time

Holly and Luis are comparing the time it takes to walk along different routes. We can subtract time, even though we can't see it.

Now You Try

This morning, Holly and Luis found a new route. It took 4 minutes less than it takes to walk along the main road. How long did the new route take?

Okay, Digit. To find out how long it takes on the quick route, we need to take 3 minutes away from the time it takes to walk along the main road.

$$10 - 3 = 7$$

It takes 7 minutes to walk through the park.

My friends all live different distances from me. Josh is 16 blocks away, Mia is 15 blocks away, and Brad is 11 blocks away.
How much farther from me does Josh live than Mia? And how much farther away from me does Mia live than Brad?

That sounds hard, but we can figure it all out using subtraction. Think about the two sets of distances and find the difference between them.

6 minutes

30

Josh lives 16 blocks away and Mia lives 15 blocks away. The difference is 16 - 15, so the answer is 1. Josh lives 1 block farther from you than Mia.

I can do the rest. To find out how much farther from me Mia lives than Brad, I need to take 11 away from 15.

$$15 - 11 = 4$$

Mia lives 4 blocks farther away than Brad.

Subtraction helps me solve problems about eggs, toys, animals, rolls of wallpaper, cookies, and lots of other things. I can write about subtraction using special signs, and I can check my answer using addition. If a number is missing in the middle of a subtraction equation, I can figure out what it is. And I can even use subtraction to find differences in time and distances!

Here are some useful subtraction words

Addition: This is a type of math that lets you join small sets together to make a bigger set.

Equation: An equation is like a sentence in math. It uses numbers and special signs instead of words.

Even numbers: Even numbers end in 2, 4, 6, 8, or 0.

Minus: This is another way of saying "take away."

Odd numbers: Odd numbers end in 1, 3, 5, 7, or 9.

Set: A set is a group of things. Subtraction is about taking a smaller set away from a bigger set.